COAL AND THE ENVIRONMENT

F O R E W O R D

This report was presented by Dr. J. Gibson at a meeting of the Resources Committee of the Institution of Environmental Sciences. The subject is of major importance. Coal's role in the energy scene, now and in the future, cannot be underestimated, but also one cannot deny the considerable impact of the industry on the environment and society. The Institution realises the positive value that can accrue from the wide dissemination of the information contained in this report and recommends its reading.

The Institution holds Dr. Gibson in high regard and it is particularly grateful for the National Coal Board's contribution to the cost of producing this publication. In seeking to sell at a subsidised price, the Institution is anxious for as wide an audience as possible. However, the views expressed in this report do not necessarily reflect the Institution's corporate attitude to all the issues involved.

British Library Cataloguing in Publication Data

Gibson, J

Coal and the environment

1. Coal mines and mining - Environmental aspects

I. Title

338.2'72 TD195.C8

ISBN 0-905927-60-5

Designed and produced by Science Reviews Ltd.

40 The Fairway Northwood Middlesex England

INSTITUTION OF ENVIRONMENTAL SCIENCES
LONDON

A series of meetings held in 1971-72 at the premises of the Royal Society, London, under the Chairmanship of the Lord Burntwood culminated in the establishment of the Institution of Environmental Sciences which subsequently became formally registered and incorporated. The Inaugural Conversazione for members of the Institution took place at the House of Lords as recently as 18th January, 1973. The Institution of Environmental Sciences is a rapidly growing non-governmental organisation concerned with the disciplinary, and in particular with the interdisciplinary approaches to environmental science. It has collective as well as individual members.

Individual Membership is open to any person without restriction of country or creed, subject to the possession of appropriate recognised qualifications and experience in matters related to the environment. Such membership is graded - Fellow, Member, Associate Member, Graduate Member and Student Member according to the Council's recognition of the individual's qualifications and expertise.

Collective Membership is available for professional and educational institutions, industrial companies and environmental consultants and agencies.

Activities The Institution has organised numerous national and international conferences and members are normally entitled to attend these at a substantially reduced fee. Proceedings of the more important conferences are published in the Institution's own international environmental journal or printed as a separate publication. Members receive throughout the year News Sheets informing them of activities both within the Institution and on the international environment scene. The News Sheet may also include short papers, book reviews and conference reports. They also receive Proceedings of the Institution which are published twice a year. These contain a number of papers of up to 7500 words.

The Institution's own Journal "Environmental Education and Information" is produced as an international quarterly and publishes new developments and research of all kinds that may have application in formal and informal education. An important feature of the journal is the Information section, providing a review of legislation, current projects and the activities of agencies and institutes; together with book reviews and journal abstracts and news of the latest developments in environmental education programmes and technology.

The Institution is also involved in various Commissions, it provides an advisory service, reduced subscription to certain Journals and is affiliated to a number of international bodies.

For full details of membership please write to:-

D.F. Potter,
Hon. Secretary,
Institution of Environmental Sciences,
14 Princes Gate,
Hyde Park,
London SW7 1PU

DR. J. GIBSON. Born 1916. Kitchener Scholar, graduated at King's College (now University of Newcastle upon Tyne) in chemistry, then research in coal constitution and carbonisation for MSc and PhD degrees; research in Northern Coke Research Laboratory and taught in University of Durham. Divisional Chief Scientist in Northern (1958) and Yorkshire (1964) NCB Divisions. Director, Coal Research Establishment 1968. Director of Coal Utilisation Research 1975; President of BCURA; member Technical Research Committees, Western European Coal Producers' Association and ECSC. President, Institute of Energy 1975-76. NCB Member for Science 1977. Chairman: IEA Working Group on Coal R & D; Executive Committee Grimethorpe PFBC Project. A Director of NCB Associated Companies. CBE 1980.

CONTENTS

5. ENVIRONMENTAL IMPLICATIONS OF COAL UTILISATION

1. INTRODUCTION

In their activities, the NCB have a firm commitment to protect, and improve the environment wherever possible. The NCB devote great attention and effort to offset any side effects from mining and using coal. The environmental issues in the coal industry are complex, and can involve a compromise with economic factors. This is not, however, unique to the coal industry; similar problems arise with every other source of energy or means of energy utilisation. Indeed, the NCB understand that it is the need to study the interaction between energy policies and the environment which led to the setting-up of the Commission on Energy and the Environment.

In common with all major industrial activities, coal production and utilisation have a considerable impact on the environment and society. This should, however, be seen in the context of the benefits which the coal industry contributes to the economy and compared with the environmental effects of alternative sources of energy and with the consequences of a failure to meet the energy needs of the nation.

The NCB appreciate the extent to which the coal industry changes the environment and has established elaborate consultative procedures to ensure that the views of those affected by mining operations and by coal utilisation are respected, as well, of course, as meeting the various statutory requirements placed on them.

The current issues of greatest concern occasioned by coal mining operations appear to be subsidence, the disturbance of land by opencast mining and the disposal of colliery spoil. The NCB recognise these as problem areas and conduct their operations in such a way as to minimise the impact and, wherever possible, to turn the problems into opportunities for positive improvement.

By planning underground operations carefully, subsidence at the surface which follows the extraction of coal underground can normally be controlled within inches

THE COALFIELDS OF GREAT BRITAIN
LOCATION OF THE MAIN CLASSES OF COAL
FIFE COALFIELD
CLACKMANNAN COALFIELD
CENTRAL COALFIELD
GLASGOW
EDINBURGH
LOTHIANS COALFIELD
DOUGLAS COALFIELD
AYR
AYRSHIRE COALFIELD
DUMFRIESSHIRE COALFIELD
CANONBIE COALFIELD
NORTHUMBERLAND COALFIELD
DURHAM COALFIELD
CUMBERLAND COALFIELD
BURNLEY
YORKSHIRE COALFIELD
LANCASHIRE COALFIELD
MANCHESTER
SHEFFIELD
IMMINGHAM BULK TERMINAL
NORTH WALES COALFIELD
WREXHAM
NOTTINGHAMSHIRE & NORTH DERBYSHIRE COALFIELD
NOTTINGHAM
SOUTH DERBYSHIRE & LEICESTERSHIRE COALFIELDS
SHROPSHIRE COALFIELD
DUDLEY
BIRMINGHAM
WARWICKSHIRE COALFIELD
SOUTH STAFFORDSHIRE COALFIELD
OXFORDSHIRE COALFIELD
OXFORD
SOUTH WALES COALFIELD
FOREST OF DEAN COALFIELD
CARDIFF
BRISTOL
BRISTOL & SOMERSET COALFIELD
LONDON
KENT COALFIELD
DOVER
SCALE OF MILES

FIGURE 1

where necessary. When damage to buildings or land does occur, the NCB aim to act quickly and fairly by carrying out repairs or providing compensation. Virtually all the claims that are made are settled locally in this way but subsidence does cause very real problems for the people affected.

The disturbance of the land caused by opencast mining is temporary and, when mining operations are finished, care is taken to restore the land and return it to agricultural use. In some cases, the land is in better condition when restored than originally. Here again, the NCB recognise that opencast mining can cause problems for those living nearby.

Methods of disposal of colliery spoil other than placing it on tips nearby have been reviewed many times by the NCB. All the other methods (for example, underground stowing or transfer to distant sites where infill is needed) are either impractical or expensive or themselves cause further environmental questions.

It should not however be thought that NCB activities necessarily lead to a net reduction in land available for other purposes. In recent years considerably more land has been released by the NCB than has been acquired for new operations. Over the last ten years, for example, the land acquired for deep-mining operations totalled 3096 hectares, whereas 8543 hectares were released.

In the future, changes in the environment will come from establishing new deep mines in areas without a previous history of coal mining. For the local communities there are, of course, immediate benefits such as higher employment and prosperity, but there are also local concerns about the direct effects of mining activities (such as those discussed above), the integration of any incoming miners into local society, and the implications for the transport infrastructure (e.g. new railways and road use). The NCB make great efforts to select sites and methods of working, waste disposal, and coal despatch to minimise disturbance. Attention is also paid to the visual aspect of the colliery workings, with trees, banks etc. being used for landscaping. In Yorkshire's Selby project, for example, 10 million tonnes of coal a year will be mined at five collieries in 25,000 hectares (100 square miles) of coalfield. However, the collieries will all be linked underground so that the entire output will be brought to the surface from two drifts at Gascoigne Wood, and the movement of coal can take place by railway with minimum disturbance to the locality. Very little minestone has to be brought to the surface and so pitheaps are avoided. Subsidence is designed to fit with a coherent solution to the neighbourhood's water drainage.

The environmental implications of coal mining are, of course, not confined to production of the coal. The way in which it is used is perhaps equally important.

The pattern of coal utilisation has changed considerably over recent years, and further changes are expected in the future, as illustrated in Table 1.

TABLE 1

Coal consumption in the U.K.

	1955 million tonnes	1977/8 million tonnes	2000 forecast million tonnes
Power stations	44	79	90
Coke ovens	27	16	20
Industry	50	9	40
Domestic/commercial	50	18	20
Railways	12	-	-
Town gas works	28	-	-
Total	211	122	170

References 1, 2, 3

At present, nearly two-thirds of the coal produced by the U.K. is burned in power stations, reflecting a general trend towards coal utilisation in large central units rather than by small consumers who have more difficulty in overcoming environmental problems. This pattern is likely to persist with the continuing growth of the use of coal in the power generation market and with a dramatic expansion of the larger scale industrial market. In the longer term, probably early in the next century, the manufacture of substitute natural gas (SNG) and liquid fuels from coal will play their part in supplying energy in the forms required for premium heating, transport, chemicals and other specialist users. The present indications are that these conversion processes are also likely to be most economic on a large scale. In many cases, the problems of controlling emissions to the environment from coal combustion and conversion processes are eased with centralised plant and the new processes are being designed with environmental considerations in the forefront.

For domestic heating, the widespread use of solid smokeless fuels and the replacement of coal, for example by natural gas and electricity, have combined to reduce smoke emissions dramatically so that dense city smogs are now events of the past. In addition, the introduction of new domestic appliances such as the "Smoke-Eaters" enables high volatile bituminous coals to be burned smokelessly and efficiently. Parallel advances in industrial appliance design and operation have reduced the emission of grit and dust in flue gases.

The combustion of coal results in the emission to the atmosphere of sulphur and nitrogen compounds and particulate material. The CEGB have undertaken detailed studies of atmospheric emissions from their plants (e.g. Ref. 4) and have demonstrated that the use of tall stacks has resulted in a considerable reduction in ground level atmospheric pollution from their stations. Similar benefits arising from the application of the tall stack policy to industry can be inferred. Although many questions about the long-range transport of atmospheric pollutants remain to be answered, recent work (Ref. 5) has led to international recognition that sulphur emissions arising from coal

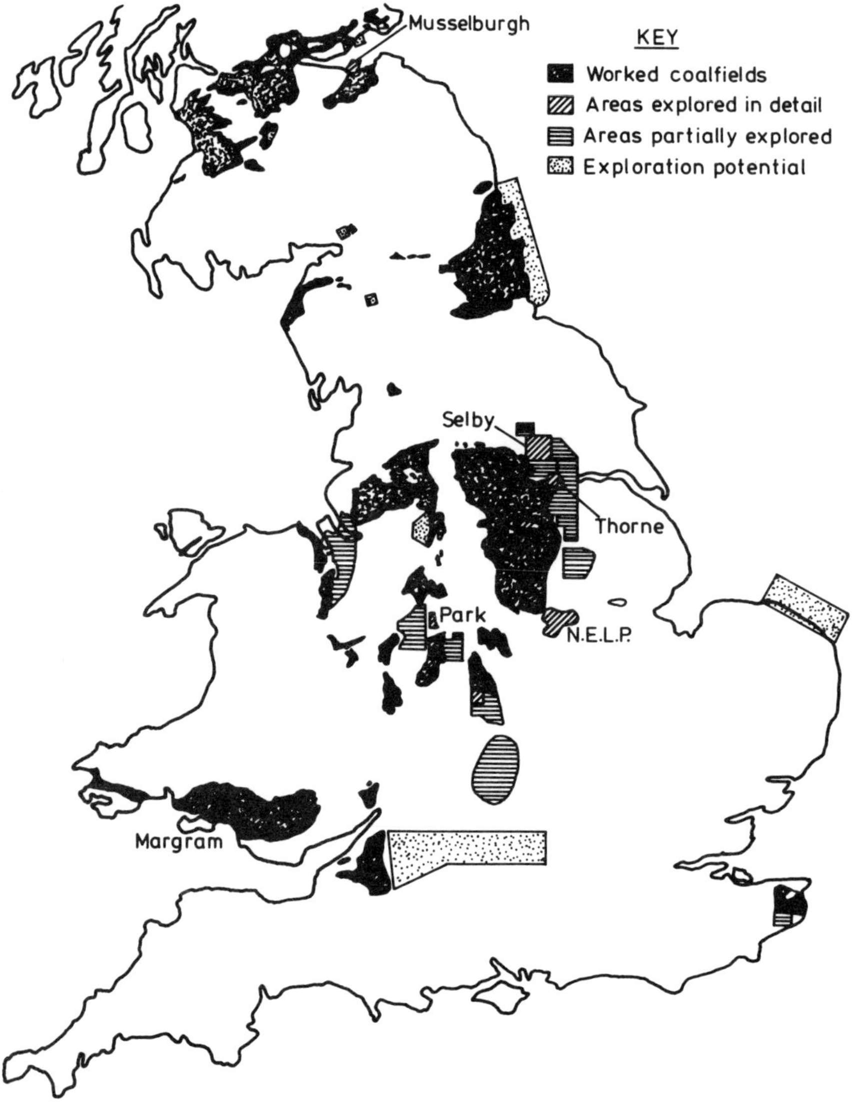

FIGURE 2. BRITISH COALFIELD EXPLORATION

combustion in the U.K. have considerably less environmental significance than had previously been suggested. The NCB, in association with other bodies such as the CEGB, will ensure that the environmental disturbance continues to be minimised.

The carbonisation of coal to make coke for metallurgical use and for the manufacture of smokeless domestic fuel is associated with special environmental problems at coke ovens and briquetting plants. Many improvements have been made both to the traditional methods at existing plants and are designed into new plants.

The introduction, at about the end of the century, of new processes for the conversion of coal will require new types of environmental control, but these are not expected to present problems different in nature from those of current oil refinery, petrochemical plants, carbonisation and power station activities.

What does perhaps tend to colour thinking about the environmental effects of coal (as opposed to other energy sources) is its long history in the U.K. and its close involvement in the dereliction associated with the industrial revolution. Coal mining, like other sources of energy, should surely be judged on its present environmental record and future plans.

2. DEEP-MINED COAL PRODUCTION

2.1 Deep mine prospecting

The locations of the established coalfields are shown on the map in Fig. 2. The potential deep-mined output of 120 million tonnes per annum by 1985 envisaged in "Plan for Coal" requires the creation of some 40 million tonnes of incremental deep-mined capacity to offset exhaustion in the older collieries and to permit some expansion. About half of the required capacity would come from the reconstruction of existing mines and the balance from new mines. To increase deep-mined output to 150 million tonnes by the end of the century will require the creation of 4 million tonnes

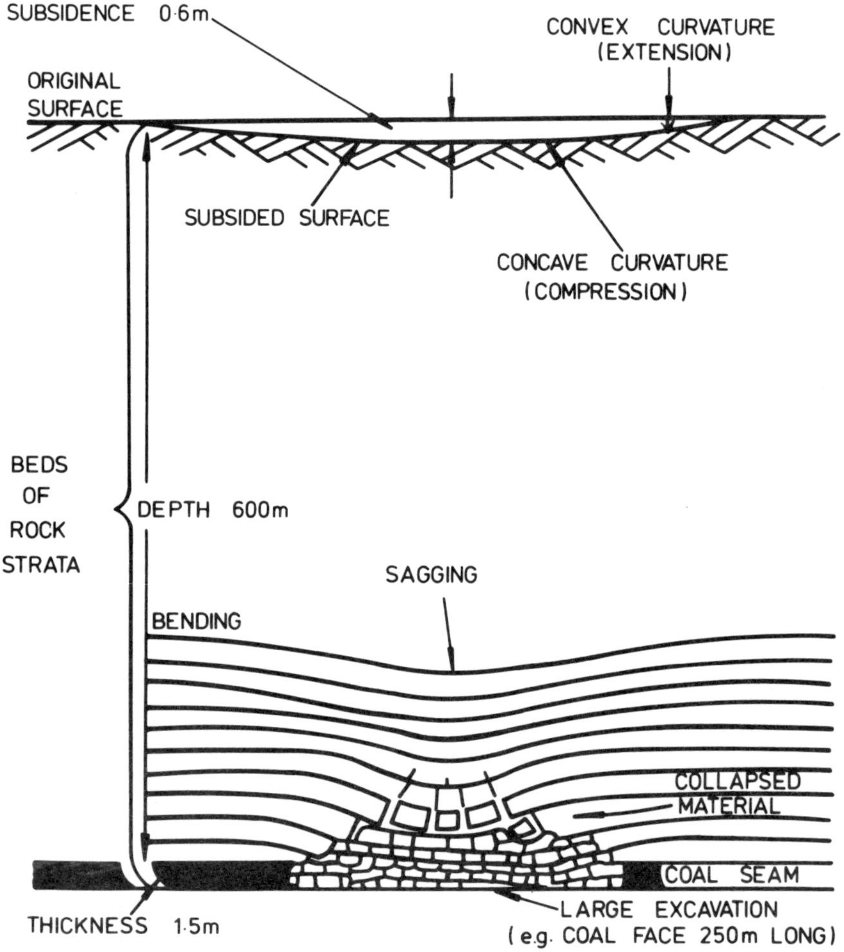

FIGURE 3. EFFECT OF A LARGE EXCAVATION (EXAGGERATED & NOT TO SCALE)

a year of new and replacement capacity over the period 1985-2000. This would mean a heavier dependence on new mines than in the current "Plan for Coal" programme and will require a continuation of the intensive coal exploration programme to supplement existing geological knowledge of the known coalfields and to prove and explore the extent of new coalfields.

The lead time for the development of new mines is about 10 years, including the time taken to obtain planning permission. Thus for a development programme to be completed by the year 2000 exploration data must be complete by 1990.

Exploration for new mines falls into four phases:

(i) The first discovery resulting from one or two boreholes sited on a geological conjecture.

(ii) Confirmation of the extent of the field by drilling further boreholes.

(iii) Detailed confirmation of the structure by a series of close spaced boreholes and seismic survey.

(iv) Further detailed information by drilling at specific points such as the site of proposed shafts, engineering and hydro-geological boreholes.

Deep boreholes are now being drilled at the rate of just over 100 a year, with about 450 km. of seismic survey. About 60% of the exploration is located in areas of possible new mine prospects.

The exploration programme for coal reserves is being carried out at locations both in and away from traditional mining areas. The locations include the area around Lichfield in Staffordshire; south and west of Coventry; between Selby and Thorne in Yorkshire; north west of York; the area around Banbury in Oxfordshire; Wiltshire and Berkshire; north Norfolk; east Cheshire; off the north-east coast; and in the Firth of Forth, (see Fig. 3). It is, of course, too early to say which of the new areas will yield viable mining propositions.

2.2 Subsidence

Almost all coal mining in the U.K. is now carried out using the mechanised longwall system whereby the coal face, which can be between 50 and 250 metres in length, is advanced by coal getting machinery which removes the coal as it travels along the face. The rate of face advance can be up to 25 metres per week and a face will normally advance for a distance of at least 1000 metres. The roof immediately behind the face is supported by hydraulic supports but behind the supports is allowed to collapse into the void left by the removal of coal.

If the area of a mining excavation is small in relation to the depth, the collapse and sagging of the overlying strata will not reach the surface. Where the excavation is more extensive, say with most longwall faces, the sagging of the strata will reach the surface and subsidence will be caused. The area affected at the surface is greater than the excavated area in the coal seam, forming a long shallow depression with a maximum lowering near the centre. The amount of lowering at any point can be calculated from the thickness of coal extracted from the seam, the area extracted and the depth of the seam. The maximum subsidence is never more than 90% of the thickness of coal extracted, and often is as low as 20%. In the area of the surface depression where the curvature is concave, the ground is compressed and in the zones where it is convex the ground is extended (see Fig. 1). There will thus be a zone of compressive strain near the centre of the depression and a zone of tensile strain on each side. For any given size of underground excavation the greater the depth of the workings, the smaller will be the amount of lowering and strain.

For example, a coal face 250 metres in length extracting 1.5 metres in thickness would, at a depth of 300 metres, cause a maximum subsidence of 1.2 metres but, at a depth of 600 metres, the same face would cause only 0.6 metres of maximum subsidence. If the thickness of extraction or the width of coal face is less, the subsidence will be smaller.

The period over which subsidence continues is governed by the depth of the seam and the speed of advance of the coal face. In general terms, subsidence at the surface starts when the coal is being worked within a distance equivalent to one half of the depth of the seam, and continues until the working passes beyond a distance of one half of the depth. For example, the period during which subsidence would occur for a face travelling at a rate of 600 metres a year at a depth of 700 metres would be about 14 months.

Except where it affects drainage gradients, lowering itself does not usually cause damage. The physical damage caused to buildings and structures is due to surface curvature or strain. The severity of damage is related to the size of the structure and the strain to which it is subjected. The strain can be measured and predicted. The NCB have evolved a damage scale classification which enables the degree of damage which a conventional structure might suffer to be assessed.

The amount of strain affecting a building will depend upon the location of the building within the subsidence zone. The maximum compression will be near the centre, the strain becoming tensional towards the flanks of the zone (see Fig. 1). Many properties within the zone will thus suffer little or negligible damage.

An investigation carried out by the NCB over a 10 year period covering the Nottinghamshire, Derbyshire and Leicestershire coalfields, showed that of the 20,000 houses in the Areas affected some 62,500 (or one in three) suffered any damage at all. Of these 62,500 the assessment of damage was as follows:

(i)	Very Slight	-	Approx. 40,000
(ii)	Slight	-	Approx. 20,000
(iii)	Appreciable	-	Approx. 2,000
(iv)	Severe	-	Approx. 250

Thus it will be seen that of the total number of houses within the area of influence the number of severely damaged houses amounted to 0.125% or 1 in 800; those appreciably damaged amount to 1%; those slightly damaged amounted to 10% and those very slightly damaged amounted to 20%.

Since most new sinkings are working seams which are deeper than average, the increased depth will, other things being equal, reduce the subsidence effects.

Various techniques have been devised and practised to reduce subsidence and damage. At the surface, precautionary measures can be incorporated in the design of new buildings by strengthening the foundations, reducing the size of individual units and allowing flexibility in service pipes, etc. With existing buildings, the measures include insertion of flexible joints in services, removing concrete against properties or in drives to reduce the build-up of stresses, cutting joints in boundary walls, shoring up large openings, etc. Underground measures involve leaving coal unworked to provide pillars of support for surface protection, although damage can occur in the regions of curvature on either side of a pillar. Research has shown that, at moderate depths, fairly wide faces may be worked without causing much subsidence. By working proportionately narrower faces in shallower seams, and by leaving pillars or unworked coal between the faces, the same effect can be achieved. By careful design of the face/pillar relationship, the amount of subsidence can be controlled. However, any system of partial extraction involves the sterilisation of a proportion of the coal reserves.

2.3 Dust and grit

Occasionally dust and grit from coal and coke stocking and handling facilities, for example from coal concentration depots, affects domestic premises. A number of dust deposition surveys have been carried out in the vicinity of plants, depots and collieries using the British Standard directional and deposit dust gauges. During normal weather conditions there are usually few problems, difficulties arising mainly when high winds occur, either during or immediately after prolonged spells of dry weather.

To alleviate such temporary nuisances, considerable attention has been paid to the facilities and methods of operation. The types of action taken are as follows:

(i) Surfacing of roads used for the transport of material.

(ii) Mechanical sweeping of open surfaces to prevent accumulations of small-sized material.

(iii) Dust suppression sprays and the enclosure of conveyors and transfer points.

(iv) The use of water sprays to moisten stocks and open surfaces.

(v) Erection of windbreaks using fences or tree planting.

2.4 Water and effluents

On average in the U.K. Coal Industry, about 3 tonnes of mine water are pumped to the surface for each tonne of deep mined coal produced. The total amount of mine water pumped in 1976/77 was 290 million m^3, and corresponds to about 9% of the total U.K. supply of water for domestic use. Most of the mine water (255 million m^3) is pumped to waste in water courses. Of this water, about 5% is suitable for domestic use after simple treatment; about 80% is relatively clean and of a quality which could be used for industrial purposes if convenient; the remaining 15% is contaminated, to some extent.

Mine waters are contaminated by suspended solids, iron compounds, chlorides and acidic sulphates. These types of contamination can occur individually or in combination. Treatment of mine waters has been undertaken where contamination has been most serious. Processes used include:

(i) Lagoons, for removing suspended solids.

(ii) Lagoons, with aeration or lime for dealing with iron compounds.

(iii) Pipelines for transporting saline waters to estuaries or to the sea.

In exceptional circumstances of adverse mine water quality (where the ground space is limited or the amenity value of the ground is high) mechanical filtration has been used instead of lagoons. This process is much more expensive than lagoon treatment.

It is expected that new regulations relating to the quality of the mine water discharged under normal working circumstances will come into effect on the 1st January, 1980. Priority attention is being given to about 20 of the 500 mine water discharges at present being made, with a further 80 or so discharges having been identified for attention later.

Under the new regulations, surface water discharges will also have to meet Water Authority requirements. In normal weather conditions, the control of surface drainage does not present a problem. However, during prolonged periods of wet weather and storm conditions, these discharges can become contaminated with material from spoil tips, colliery yards and possibly also from coal stockpiles. The extent of this problem and methods of alleviating it are being investigated jointly by the NCB and the Water Authorities.

Occasionally, iron compounds occur in seepages from spoil tips and coal stockpiles; where necessary, these are controlled by treatment.

Where practicable, coal preparation plants have closed circuits and produce virtually no effluent. Where effluents are produced, these are normally discharged through the same consented outlet as the mine water and other discharges from the mine. Suspended solids are removed by flocculation and sedimentation and the effluent is generally of good quality by the time it reaches the discharge point. Effluent problems sometimes arise from a mechanical failure of the plant or human error, but prompt action is taken to rectify these situations.

The NCB carry out a large programme of monitoring the quantity and quality of the effluents. The NCB understand that the Water Authorities do not currently regard coal-mining as a major contributor to river pollution, which is a dramatic change from what would have been said 20 or 30 years ago.

2.5 Noise

Where there are difficulties over noise at particular collieries these can generally be traced to individual items of plant such as fans, compressors, screens and wash boxes at coal preparation plants, railway sidings, road vehicles and audible warnings associated with safety during the manoeuvring of vehicles. The NCB make all reasonable efforts to identify and deal with such sources of noise.

When planning new collieries, the NCB are concerned to ensure that environmental noise is kept to a minimum and, indeed, Local Authorities need to be satisfied on this aspect. At Selby, agreements were reached with the local authorities on the siting of major items of colliery plant to ensure minimum disturbance to the neighbourhood. In the North East Leicestershire coalfield the same procedure is being planned.

Borehole drilling is sometimes necessary in built-up areas. Even though drilling at a particular location is of short duration, appropriate noise abatement measures are taken by, for example, enclosing the drill's superstructure, using silencers on drilling motors and, in some cases, by surrounding the drilling rig with sound absorbent materials.

Coal concentration depots are usually part of existing railway marshalling yards and noise is generated by the normal railway operations of shunting in sidings and wagon emptying. Where appropriate, the NCB take measures to protect the surrounding properties from noise by enclosing wagon emptying operations; similar protective arrangements are applied to coal screening plants.

In general there are few complaints about noise to NCB.

2.6 Spoil disposal and utilisation

For every tonne of saleable coal produced, on average about half a tonne of dirt is discarded. The total quantity of dirt to be disposed is therefore about 50 million tonnes per annum and requires approximately 200 hectares of new land for tipping each year. (This requirement is not, of course, a net addition to NCB land use - indeed, the NCB are currently returning to other uses nearly 3 hectares of land for each new one they employ). Existing tips have been estimated to occupy more than 11,000 hectares and to contain 2,500 m million tonnes of material.

Many tips were built in the past with little regard, by present standards, for environmental considerations. This resulted in an adverse attitude by planning authorities, environmentalists and the general public. Over the years, however, much experience has been gained in the building of tips. Exacting technical specifications are now observed from both safety and environmental viewpoints.

With the use of modern civil engineering equipment, the spoil is now spread in layers and the tips are profiled and shaped as they are constructed. The contours are landscaped to blend with the lie of the land and graded to take account of the drainage necessary to ensure the safety of the finished tip. Progressive restoration with top soiling and seeding may be possible whilst the tip is in use, and virtually eliminates the risk of tip fires.

On economic grounds, it is preferable for a tip to be adjacent to the mine if a site is available. This becomes progressively more difficult as a colliery's life is extended. There is, therefore, a constant search for new areas of land where rights and consents can be acquired. Tipping at sites remote from the mine has disadvantages of cost, difficulties of site control, the possible unreliability of the transport system and adverse environmental effect generated "en route" by long distance transport (particularly if road transport is used)

The underground stowing of waste was carried out in this country on a limited scale some years ago but declined with the introduction of modern mining methods which entail rapidly advancing highly mechanised faces. There are also serious safety and technical objections to stowing as the system is associated with high dust and noise levels at the coal face. The preparation and handling of dirt at the surface and transporting the dirt back underground would require very extensive modifications to existing facilities, both on the surface and underground. In addition, support systems at the coal face would need to undergo a major re-design to ensure safe working conditions.

The Minestone Executive of the NCB is responsible for the commercial utilisation of colliery spoil and for finding new uses for this "minestone". This includes "burnt shale", the material from older tips where combustion has occurred. Approximately 7 million tonnes per year of these materials have been removed for disposal or sale, mostly for road construction, bulk fill purposes and land reclamation. Spoil can also be upgraded by heat treatment into a lightweight aggregate for making concrete blocks.

2.7 Land requirements

The total area of land occupied for colliery operational purposes, including spoil heaps, is about 23,000 hectares. In addition, 450 hectares are occupied by NCB (Coal Products) Ltd. and 240 hectares by NCB (Ancillaries) Ltd. Coal Industry Estates Ltd. is responsible for the management of non-operational land held by the NCB. This comprises approximately 46,000 hectares of farms, 2,200 hectares of woodlands and 15,000 hectares of other estates.

During the last 10 years, about 3,100 hectares of land has been acquired for colliery operational use for new mines, extensions to existing collieries, new tips and tip extensions, coal stocking sites, etc. This land includes agricultural land already in the ownership of the NCB but taken out of farm tenancies and put to operational use. The figure does not include land for opencast activities.

Over the same period, about 3,000 hectares of former operational land, such as the sites of closed collieries, coke-works, etc., has been disposed of by sale or lease for public and private development. In addition, about 5,500 hectares of former operational land (including disused tips) had been sold to County and District Councils for derelict land reclamation schemes. Land reclaimed by these schemes is normally used for public open space, afforestation or agriculture. These figures do not include restored opencast land.

The net land transactions of the NCB over the last 10 years have therefore resulted in the return to beneficial public use of approximately three times as much land as has been acquired for all deep mine operations.

2.8 Employment effects

Coal mining, although it involves increasing use of capital, remains relatively labour-intensive. For example, the new coal mines at Selby are expected to employ about 4000 mineworkers, and the three proposed for North East Leicestershire slightly fewer. The requirements are, of course, predominantly for men, since there are statutory requirements preventing the employment of women underground, except in limited circumstances.

Whilst many of the men employed at a new "greenfield" colliery will usually come from other coalfields, there will be the opportunity for significant numbers of men already living in the neighbourhood of the new sites to obtain mining employment. Further, the arrival of such a large industrial unit will generally bring with it "multiplier effects" for the local community, such as services required by the mines and those who work at them.

In some cases, the sinking of a new mine will provide employment opportunities, sometimes within daily travelling distance, for men at existing mines where the reserves are approaching exhaustion. This is, for example, a significant point in relation to the development of the North East Leicestershire coalfield.

3. OPENCAST COAL PRODUCTION

The reasons for the NCB engaging in opencast production are discussed below. The NCB activities are organised by their Opencast Executive. Whilst there is considerable literature on deep-mining techniques, and we therefore did not give a detailed explanation of these in Section 2, it seemed helpful to provide a short description of opencast techniques as well as some discussion of their environmental implications.

3.1 The need for opencast coal

In 1974, the NCB prepared their "Plan for Coal" and a Tripartite Committee, consisting of the Government, the NCB and the mining industry Unions, was set up "to consider and advise on the contribution which coal can best make to the country's energy requirements and the steps needed to secure that contribution". Both the interim and final reports of this Coal Industry Examination (Ref. 23, 24) which were produced and published in 1974, confirmed the NCB's plans, which included increasing opencast production to at least 15 million tonnes per year.

The Government's endorsement was subsequently reaffirmed in "Coal for the Future", published in 1977 by the Department of Energy (Ref. 25). Paragraph 26 stated that "Opencast coal has a number of advantages. It is usually of high quality and is often needed for blending with poorer deepmined coals to help them meet customer specifications. It saves imports by meeting something like half of our essential anthracite needs and much of our needs for prime quality coking coal which cannot otherwise be met. It is cheap to produce and helps towards the overall viability of the industry".

3.2 Opencast prospecting

Two main sources of information are available to the NCB to assist in the prospecting of coal suitable for working by opencast methods:

(i) Abandoned plans for any deep mine working which may have been proposed in the area are studied;

(ii) A systematic and extensive drilling operation mounted in the area.

At present the NCB are operating, through their contractors, some 70 drilling rigs throughout the country with 130 geologists evaluating the results. Last year some 22,500 boreholes were drilled to an average depth of 35 m. Coal is being proved at a faster rate than it is currently being worked, thus giving some measure of flexibility in the eventual sites for which applications for authorisations are subsequently made.

3.3 Overburden excavation

Excavation begins with topsoil stripping followed by subsoil stripping. The soils are stocked in mounds so as to be available for eventual site restoration. Topsoil mounds act as baffle embankments where necessary on the site perimeter and help considerably to reduce noise outside the site. They are seeded to grass and act as a screen for the working operations.

All available topsoil is recovered together with 600 to 900 millimetres of subsoil. If there is likely to be a deficiency in soil thicknesses, the soil mounds are supplemented by soil-making material recovered from the overburden at lower depths during excavation. This soil-making material is identified during prospecting drilling against the eventuality of a shortage of topsoil/or subsoil. Nowadays, soil stripping is usually done using motorised scrapers.

The choice of main excavating plant is related to the geology of the site. Dragline working is favoured wherever possible because it is the cheapest method of moving overburden in the U.K. The largest dragline in use in this country has a capacity of 49 m^3, and 10 draglines currently have a capacity in excess of 11 m^3. Where it is uneconomic or impractical to use draglines, mining shovels (up to 10.5

m^3 capacity), hydraulic shovels (up to 7.5 m^3 capacity) and backhoes (up to 10 m^3 capacity) are used. Dumptrucks are used for hauling overburden within the site. The largest has a capacity of 90 tonnes, although the most common class carries 45 tonnes.

Production contracts define the 'provisional ratio' as the ratio of the volume of overburden vertically above the coal to the volume of the coal itself, all in situ. At present, the average provisional ratio is about 15.1 and the maximum provisional ratio 35:1. The 'working ratio' of overburden to coal includes an allowance for the slope needed from the bottom of the excavations and is therefore substantially higher than the provisional ratio. The average depth of working is 50 m and the maximum depth planned at present will be 214 m at the Westfield site in Scotland.

3.4 Coal excavation

The excavation used for producing coal at the seams bared by the main overburden excavating machines are comparatively small rope or hydraulic crowd shovels of the order of 1 m^3 capacity. These load directly into coal lorries at the face. The coal lorries may be normal road vehicles taxed for hauling on the public highway, or may be rather larger if hauling over a private, internal coal haul road direct into an Opencast Executive disposal point.

3.5 Public highways

To avoid mud or coal being deposited on public highways, washing facilities are installed before coal production begins. Loading is controlled so that there is a generous freeboard between the top of the coal and the top of the coal-lorry body sides and tailboard to avoid spillage. Where necessary (e.g.in disposal points) road brushes are used. Points of access on to the public highways from the site are agreed with the Highway Authority and are designed with vision splays approved by the Authority.

3.6 Blasting

Control of blasting is based on an initial rest blast and vibration investigation carried out by explosives experts who are independent of the NCB. The criterion used for control is the peak particle velocity (ppv) of the ground motion. Blasting procedures and schedules are generally designed to limit the peak particle velocity to a maximum value of 12 mm/s.

The limiting level of 12 mm/s is considerably less than the value of 50 mm/s generally regarded as a safe level of vibration below which structural damage is unlikely in well-founded and well-constructed buildings. Although this level of blasting ensures that there is virtually no possibility of strutural damage, people can still be conscious of blasting vibration. Where necessary, therefore, blasting is reduced still further, subject to what is practicable and desirable. The times of blasting are strictly controlled and a complete record is kept of every shot fired.

3.7 Dust and grit

The main sources of dust on opencast sites are the dirt haul roads within the excavation areas. This problem is dealt with by providing water bowsers to keep haul roads damp at all times during dry weather. Other dust prevention measures include fitting blasthole drills with dust collection equipment and plant and vehicles with vertical exhausts.

Special arrangements are often made for dumptruck tipping on overburden tips, or for dragline casting, when high winds blowing across the site pose problems of dust control.

3.8 Water and effluents

Clean water entering the site is diverted around the excavation areas to avoid contamination. Site effluents are subject to consent by Water Authorities who stipulate consent conditions for a number of parameters, the most important of which are the pH value (usually required to be within a range of values from 6 to 8), the phenol content (usually limited to a maximum value of 5 milligramme per litre), and suspended solids. Standards set for suspended solids have varied widely; in the view of the NCB, a practicable limiting value is 100 milligramme per litre except in special circumstances. The NCB have not always been able to comply with consent conditions for suspended solids by using settlement lagoons alone, and the addition of flocculants has been necessary.

Flooding has sometimes occurred during the summer months, immediately following contractual restoration, when the surface has been restored but vegetation has not been re-established and sudden thunderstorms with rainfall of high intensities have passed over the site. Ditches and culverts are now designed for the worst possible conditions during the contractual restoration phase. Watercourses downstream are assessed and any necessary stream improvements carried out. Ditch erosion is overcome by stone pitching, forming cascades or weirs, or, if required, placing stone-filled gabions or using concrete channels.

3.9 Noise

The reduction of noise to a minimum is considered to be one of the most important aspects of site control. Precautionary measures are taken for all plant and vehicles before they are put into service. After such treatment, measurements show that most items of excavating plant and dumptrucks are within a maximum sound level of 90 dB(A) at 7 metres.

However, in the NCB's view, the main criterion for assessing the environmental effect of noise from opencast working should not be the noise generated by a single item of plant, but the combined effect of all items of plant on noise levels in the vicinity of the site. Baffle embankments have already been mentioned; in addition, over part of the haulage cycle, the excavations themselves act as baffles and effect a considerable attenuation of noise levels.

The NCB have been accumulating noise records around the boundaries of opencast sites for 6 years. The noise levels on site boundaries only exceed 70 dB(A) for short periods and, generally, noise levels are much lower than this during general site working. Noise control is now specified in terms of the equivalent continuous sound level, Leq. On the basis of the more limited data available on Leq. noise levels, it is possible to limit general site noise at site boundaries, or at the nearest noise-sensitive property depending on site topography, to a maximum level of 65 dB(A) Leq during daytime hours. Throughout the life of the site, noise levels at any one location adjacent to the site boundary are considerably lower than this value for most of the overall period. Leq levels are measured over a shift period of 8 hours.

Exceptions to the limiting level of 65 dB(A) Leq are necessary for motorised scrapers working close to the site boundary on soil stripping and the construction and removal of soil mounds. It is also usual to exempt blasting noise, normal coal road traffic, and certain other subsidiary operations of a civil engineering nature which occur only occasionally. For night work, the NCB undertake to restrict noise levels at site boundaries or at noise sensitive properties to a maximum level of 55 dB(A) Leq.

Plant yards and service areas are screened from the public and the standards of site offices and buildings have been improved in recent years.

3.10 Change of surface features

Many proposed opencast coal sites contain evidence of former industrial dereliction, ranging from old shale heaps to agricultural land the surface of which has been affected by subsidence. During the course of winning the coal, this dereliction is cleared away and the land eventually restored to a more beneficial use.

Most of the land required for opencast mining is, however, agricultural land. The occupier receives full compensation for the loss of income from the land while the site is being worked for coal. Where only part of the holding is required, it is often possible for the occupier to carry additional livestock by buying in more foodstuffs.

Before opencast coal mining can begin, detailed discussions take place with the statutory undertakers whose services may cross the site. Wherever possible these are diverted permanently, but sometimes it is preferable to arrange for temporary diversions. This particularly applies to public roads and footpaths.

After preliminary work on a site has been completed, the topsoil and subsoil are removed by scrapers and stacked separately. On small sites this may all be done at the same time, but on large sites stripping will be progressive in advance of coaling operations. Vegetation is left undisturbed for as long as possible and land clearance and soil stripping kept only as far in advance of main excavation as is necessary to maintain the site working programme.

The contractor then starts to dig an initial void or box cut with this overburden being placed in a mound in the most suitable location. The coal is then removed and transported to the disposal point whilst the overburden from the adjacent area is dug with a dragline and cast into the previous void. This sequence continues until all the coal has been removed and the final void isthen filled with the overburden from the original box cut. As far as is practicable, back-filling of the voids and restoration are progressive so that the land can be reinstated and grass cover re-established as soon as possible after coaling.

On deep or complex sites, and particularly where a number of seams are present, the simple dragline technique has to be supplemented with additional earth moving equipment to achieve overburden reduction and removal. The seams are then worked in a series of benches or terraces with face shovels and dump trucks being utilised to remove some of the overburden. This often results in a large overburden mound being formed which may then have to remain in the same location until coaling has been completed.

Overburden tips and soil mounds are shaped and graded to make them as unobtrusive as possible. The heights of tips and mounds are often restricted by local agreement and care is exercised to ensure that, as far as is practicable, overburden tips do not protrude above the skyline when viewed from certain critical vantage points. Experimental work is being carried out to spread grass seed so that large mounds can become less obtrusive.

3.11 Land requirements and reclamation

To sustain the target annual production of 15 million tonnes of opencast coal, some 2,000 hectares of land must be brought into coal production every year. The average site produces coal for about four years and then undergoes a five year restoration period. This means that some 8,000 hectares are required for coal winning purposes at any one time whilst a further 10,000 hectares are undergoing restoration.

When coaling operations have been completed, the overburden is levelled by bulldozers and graders to the required contours. The subsoil and topsoil are then carefully re-spread with the separate layers rooted to avoid panning and to remove large stones. Officials from the Ministry of Agriculture Fisheries and Food are present at each of these stages. When topsoiling has been satisfactorily completed the site is accepted by the Ministry of Agriculture and begins a five-year period of agricultural rehabilitation under their supervision at the expense of the NCB.

The work of restoration falls into two main categories:

(i) The restoration of the farm fixed equipment such as fences, ditches, hedges, trees and woodlands. This work is usually carried out on behalf of the NCB by specialised contractors.

(ii) The rehabilitation of the land by careful management of cultivation, manuring, cropping, control of grassing, the provision of water to the fields and the design and installation of a permanent under-drainage system. The Ministry of Agriculture, as agents for the NCB, undertake this work, using either agricultural contractors or the occupiers of the land.

Restoring the farm fixed equipment, and farming the land, go on concurrently throughout the five year period, the object being to re-establish a good soil structure. Some grassing is permitted during the summer months but, in the wetter parts of the country, the grass is generally cut and conserved as silage. Some woodland is re-planted either in the form of shelter belts or spinneys in the corners of fields. In South Wales, much of the restored land is unsuitable for intensive agriculture, especially the hill and mountain slopes or areas of peat. These areas are usually restored as forest with the land restoration work being undertaken by the NCB's site contractor but the actual planning, management and maintenance of the restored woodland is carried out by the Forestry Commission acting as agents for the NCB.

During the course of restoration, water courses will be put back in agreement with the appropriate Water Authority. Footpaths and public roads will also be reinstated on their original alignments to a condition "not less serviceable than at the commencement of operations". If farm buildings have been destroyed in order to dig the coal, new buildings will be erected. As these will be required before the land has had sufficient time to become stable, these are usually located on a part of the site which has been used for soil storage and has therefore not been disturbed.

After the land has received its five-year treatment under the supervision of the Ministry of Agriculture it must continue to receive good farm management if it is to recover to its original productivity in about 10 to 15 years time.

Care must therefore be taken to avoid poaching and trampling by livestock in wet weather and additional quantities of fertilisers may be required.

Production of coal by opencast methods was started during the last war when only some of the topsoil was stripped and mounded to restore the land. This led to many complaints and eventually resulted in the Ministry of Agriculture drawing up an improved code of restoration in 1951. In the late 1950's some research work was carried out by the Ministry of Agriculture, Fisheries and Food (Ref. 6).

Plans for further research were, however, abandoned in the early 1960's and the 1951 restoration code still forms the basis of present day practice, although many improvements have been made in both agricultural technology and the type of equipment being used in the opencast coal mining operations. Following the expansion of opencast production envisaged in "Plan for Coal" (Ref. 7), research work has been re-started both with the Ministry of Agriculture and other external organisations. These include the Soil Association in conjunction with the Soil Science Department of the University of Wales, the Institute of Terrestrial Ecology at Bangor and Newcastle University in association with their Cockle Park Experimental Farm.

4. BACKGROUND TO COAL UTILISATION TECHNOLOGIES

Sections 4.1 to 4.6 present a brief summary of the main coal utilisation technologies (both in current use and under development).

4.1 Power generation

At present, most coal used for electricity generation is burned in pulverised fuel (PF) combustion systems. In this system, the coal is pulverised so that the maximum particle size is 0.1 mm and burned in a high temperature flame. Heat for steam raising is extracted from the gases which are passed through electrostatic precipitators to remove particulate matter. The gases are then emitted via the tall stacks which CEGB have shown to be effective in dispersing the products of combustion and minimising ground level concentrations of pollutants (Ref. 4).

Sulphur emissions from existing power stations could, in principle, be reduced by about 90% using stack gas scrubbing systems. These have been demonstrated to be technically feasible on several large-scale power stations in the USA and elsewhere. There are, however, significant economic penalties; the capital cost of the station is increased by about 20%, and 12% more coal is consumed for the same electrical output (Ref. 8). It is important to note that, while the increase in the cost of electricity can be quantified, the justification for incurring these costs still rests on widely differing views of the environmental effects of sulphur emissions.

Further, the NCB are actively pursuing the development of two new power generation technologies. These are fluidised bed combustion (FBC) and low calorific value gasification (LCVG), both of which could lead to reduced atmospheric emissions.

Fluidised combustion involves burning coal in a bed of particles through which the air for combustion is passed. The particle size of the bed material (0.5 mm to 1mm) and velocity of the air 1 m/s to 3m/s) are chosen so that the bed is in a fluidised state and resembles a boiling liquid. Coal combustion in a fluidised bed takes place at the relatively low temperature of about 850°C, compared with about 1600°C in a PF furnace. A consequence of this is that nitrogen oxides (NO_x) are not produced from atmospheric nitrogen in FBC. Although NO_x are still formed from the organic nitrogen in the coal, the overall amount is reduced. Up to 90% of the sulphur in the coal can be retained by adding an acceptor (limestone or dolomite) to the fluidised bed. The sulphur is captured as calcium sulphate which is disposed with the ash and unconverted acceptor. Fluidised bed combustion can take place at either atmospheric pressure or at elevated pressure. In the latter case, the use of both gas and steam turbines (combined cycle operation) results in an increase in the electricity generation efficiency and corresponds to a reduction in coal requirements of 5% to 10% for a given output. Operation at elevated pressure also enhances some of the advantages of atmospheric pressure operation; sulphur retention occurs more readily and nitrogen oxide formation is further reduced.

The NCB have pioneered the development of fluidised bed combustion at both atmospheric pressure and elevated pressure and are acknowledged world leaders in this field. For power generation applications, effort is directed towards pressurised fluidised bed combustion (PFBC). The 5 MW(th) pilot plant at the NCB's Coal Utilisation Research Laboratories (CURL) at Leatherhead is the largest operational PFBC in the world. By the end of 1980, operation of an 80 MW(th) experimental facility at the NCB's Grimethorpe Colliery is scheduled to begin. This project is jointly funded by the U.K., West Germany and the U.S.A. under the aegis of the International Energy Agency, with the NCB acting as operating agents. The objective of the project is to obtain basic design data and operating experience of PFBC over a wide range of conditions.

Pressurised fluidised combustion is also the subject of a proposal submitted by the NCB to the Government, in collaboration with Babcock & Wilcox with the support of the CEGB. This project arose from a recommendation of the Coal Industry Tripartite Group, April 1978 (Ref. 9).

An alternative method of improving the efficiency of power generation is to convert coal into a clean fuel gas and use this in an advanced combined cycle system. A potential economic advantage over FBC is the ability to use higher combustion temperatures and advanced gas turbines to give a higher cycle efficiency. These gains may or may not be enough to offset the loss in efficiency during the gasification state. Emissions of sulphur from a gasification combined cycle plant can be reduced to a low level (less than 1% of the sulphur in the coal) by cooling and chemically processing the gas. If required, the sulphur can be recovered in solid form for ease of disposal. Some of the organic nitrogen in the coal is also recovered (as ammonia) in such processes. However, the formation of NO_x from atmospheric nitrogen is a potential problem as may be the case with some liquid fuelled industrial gas turbine combustion systems. Some gasification/gas clean-up systems produce considerable quantities of waste water, but this can be purified by techniques based on those in the oil industry.

The NCB, supported by the ECSC, have a research programme to develop an efficient gasification system for combined cycle power generation. A joint study between the NCB and CEGB has been initiated to define proposals to the Government for a pilot plant project in this field. This also follows from a recommendation by the Coal Industry Tripartite Group's report of April 1978 (Ref. 9).

The timescale required for the introduction of new power generation technologies is such that the first commercial power stations based on fluidised bed combustion or low calorific value gasification are unlikely to be operational in the U.K. until much before the turn of the century.

Several other advanced power generation systems have been proposed, including magneto-hydrodynamics (MHD). The NCB have no research on this topic, and regard the implementation of this difficult technology as being on a longer timescale than the technologies discussed above. There appear at present to be no significant environmental advantages for MHD, compared with other advanced, high efficiency systems.

4.2 Industrial sector

The industrial and quasi-domestic market covers an extremely wide range of boiler output requirements ranging from the need to heat a school or office to the supply of heating, process steam and possibly electricity to a large industrial complex. The boilers used for each application vary in design and cover different ranges of output, but above 2 mW(th) they are of two main types, shell boilers and water tube boilers.

The pattern of boiler use in the U.K. is classified in Table 2. As implied by the table, conventional coal-fired boilers are considerably larger than oil-fired or gas-fired units of equivalent output. This disadvantage is compounded by the need to adopt less favourable construction technique. Furthermore, the cost of the ancillary equipment (including coal and ash handling), which accounts for approximately half of the total cost of a coal-fired system, is also considerably greater than for oil-fired and gas-fired systems. The above factors combine to make the cost of a conventional coal-fired boiler system two to three times higher than the equivalent oil or gas-fired system throughout the range of industrial boiler sizes.

TABLE 2

Industrial boiler classification

Size range	Coal	Oil/Gas
2 MW(th) to 5 MW(th)	Horizontal or vertical shell boiler with single combustion chamber	
5 MW(th) to 10 MW(th)	Horizontal shell boiler. Two combustion chambers.	Horizontal shell boiler. Single combustion chamber.
10 MW(th) to 20 MW(th)	Water tube boiler. Factory fabricated.	Horizontal shell boiler. Two combustion chambers.
20 MW(th) to 40 MW(th)	Water tube boilers. Site fabricated.	Water tube boilers. Factory fabricated.
Above 40 MW(th)	Water tube boilers site fabricated.	

With rising oil prices and less gas available for non-premium uses, the NCB expect coal to regain a substantial share of the industrial bulk heating market. Coal is most competitive at industrial sites where coal-fired boilers have already been installed and handling facilities for coal and ash already exist, although important opportunities also exist for new installations and 'retro-fitting', i.e. changing from oil-firing to coal firing. In order to strengthen the position of coal in this sector, the NCB have an active development programme to apply fluidised bed combustion in the industrial market. Industrial fluidised bed combustors can be considered in two categories, according to the size grading of the coal used:

(i) The coal is dried and crushed to a top size of 1 to 5 mm before feeding to the fluidised bed.

(ii) The coal is fed to the combustor as delivered, generally washed and with a top size of about 25 mm. It has been found that, with an appropriate choice of bed particle size and fluidising velocity, the large coal particles can float in the bed. The use of uncrushed coal permits compact combustion systems based on shallow beds and high air velocities to be employed, and is particularly suitable for the smaller sized boiler designs.

Both types of industrial boiler are under active development in the U.K. A 3 m x 3 m water tube boiler burning crushed coal (type (i)) has been built at Renfrew by Babcock & Wilcox and Combustion Systems Limited. The output from the unit is approximately 10 MW(th) steam. When tests are complete, it is planned to construct 5 prototype commercial boilers (3 in Europe and 2 in the U.S.A.). The design concepts in which present industrial gradings of coal are used (type (ii)) have been pioneered by the NCB. Several prototype boilers of 2 to 30 MW(th) are being installed and tested. The principle has also been applied to drying furnaces of 5 MW(th).

Inevitably, an increase in the industrial market for coal will bring an increase in coal transport. Already most of the industrial market is supplied with washed coal to reduce the quantity transported and ash disposal problems at the point of consumption. Improvements in boiler design have considerably reduced grit and dust emissions and, in general, the sulphur content of U.K. industrial coals is less than that of fuel oil. Research is in hand to apply to the industrial boiler scale the techniques of sulphur retention investigated for the larger power station boilers, although it is considered probable that in the U.K. the present policy for chimney height control will remain the preferred anti-pollution policy for industrial combustion sources for many years to come.

4.3 Domestic sector

The most noticeable environmental effect of domestic coal consumption in the past has been smoke emission, and the resultant smogs. These have now been virtually eliminated in the majority of urban areas by the widespread use of smokeless solid fuels and by the replacement of coal in the domestic sector, for example, by natural gas and electricity.

Smoke production results from unburned volatile matter leaving the fire, and is therefore an indication of inefficiency in combustion.

Three main methods of avoiding this with solid fuel exist:

(i) Burning anthracite.
Because anthracite is low in volatile matter, it is a natural smokeless fuel.

(ii) Burning manufactured smokeless fuels.
The component of the volatile matter responsible for smoke formation can be removed by heating coal in the absence of oxygen. The resulting smokeless fuel is solid either as coke or briquettes.

(iii) Use of "Smoke-Eater" appliances.

These are able to burn bituminous coal smokelessly. The principle of operation is to pass the combustion gases through a hot oxygen-rich region before they enter the flue, thereby ensuring virtually complete combustion of any unburned volatile matter.

The combustion of coal for domestic purposes also results in emissions such as sulphur dioxide, nitrogen oxides and relatively coarse particulate matter (grit or dust).

The total consumption of solid fuel in the domestic and commercial sectors has been falling over recent years and was 17.5 million tonnes in 1977/78 (see Table 3).

TABLE 3

Disposals of solid fuels in the domestic and commercial sectors for 1977

Market	Consumption (million tonnes)
Smokeless fuels	
Mainly open fires:	
Homefire briquettes	0.227
Non-NCB manufactured(includes Coalite & Rexco)	1.411
Mainly roomheater and boiler fuels:	
Anthracite (includes imports & 0.121 million tonnes to miners)	1.384
Dry steam coal	0.240
Phurnacite	0.737
Hard coke	0.177
Other:	
House coal (mainly open fires)	7.770
Miner's coal (excluding anthracite)	1.749
Commercial sector	
Coal to national & local authorities, agriculture and miscellaneous	2.060
Coke and breeze as above	0.510
TOTAL	17.543

Reference 1.

Improvements in the efficiency of coal consumption by the consumer can be achieved by the replacement of open fires with closed roomheaters or boilers (see Table 4). At around the turn of the century, coal is also expected to make a substantial contribution to domestic heating requirements through the extension of district heating schemes and the manufacture of substitute natural gas (SNG).

TABLE 4

Solid fuel appliance efficiencies

(figures as per cent)

1. Open fire without boiler						
Burning coal	26	Chimney gain	4	Total	30
Burning Smokeless fuel	37	Chimney gain	3	Total	40
2. Open fire with domestic boiler						
Burning coal	37	Chimney gain	3	Total	40
Burning Smokeless fuel	47	Chimney gain	3	Total	50
3. Open fire with high output boiler						
Burning coal	47	Chimney gain	3	Total	50
Burning Smokeless fuel	57	Chimney gain	3	Total	60
4. Exempted Openable roomheater (Rayburn Prince 76)						
Burning coal (smokelessly)						
Open	50	Chimney gain	5	Total	55
Closed	55	Chimney gain	5	Total	60
5. Roomheater without boiler						
Burning smokeless fuel	60	Chimney gain	5	Total	65
6. Roomheater with boiler						
Burning Coal (Smoke Eater)	65	Chimney gain	5	Total	70
Burning Smokeless fuel	70	Chimney gain	5	Total	75
Independent boiler						
Burning smokeless fuel	65			Total	65
Burning small Anthracite	75			Total	75

Notes:

1. The installation of a throat restrictor for open fires without a boiler will increase space heating gains from the appliance by some 2 to 3%.
2. Some open fires with high output back boilers have efficiencies exceeding 70% but figures shown are typical average operating efficiencies and assuming some throat restriction.
3. Exempted openable roomheaters give a higher flue gas temperature than open fires with a consequent increase in chimney gain.
4. No credit is given to independent boilers for casing of chimney heat gain, although in fact a typical installation with a cast iron flue pipe connection may provide up to about 1 kW of extra warmth to the kitchen.

Reference 10.

4.4 Metallurgical coke and smokeless fuel

Coke is made by heating crushed coal in the absence of air in slot type ovens lined with refractory brick. These ovens, which vary in size from 8 to 15 m long, 3 to 8 m high and 350 to 600 mm wide, are arranged in batteries of 10 to 100 ovens. Heating of the ovens is by the combustion of gas produced in the carbonisation process in flues passing through the brickwork separating the ovens to maintain the oven walls at temperature of 110 to 1350°C.

Two types of coke are manufactured in the United Kingdom at the present time; domestic coke for use mainly in roomheaters, closed stoves and boilers, and metallurgical coke for foundry and blast furnace use. Each has its own specification and is produced under different conditions. Domestic coke is the most reactive in order that it should ignite readily and continue to burn when banked overnight; it is relatively small in size and is not particularly strong. Foundry coke must be in large pieces of high strength and the amount of ash, moisture, sulphur, etc. must not exceed certain limits. Blast furnace coke must also have high strength even at the high temperatures of the furnace and the size, reactivity and purity of the coke must satisfy strict conditions. These different qualities of coke are obtained by careful selection of carbonising temperature and time, and the the type of coal or blend of coal used. The dimensions of the ovens, particularly their width, also have an effect.

Not all coals can be used for making coke; only those which possess particular plastic and contraction properties when heated in the absence of air are suitable. In the past, metallurgical cokes were produced only from coals of prime coking quality, the U.K. reserves of which are now depleted. Although research in coal blending has given some alleviation, the pressure on coking coal supplies has been increased by developments in blast furnace technology which have resulted in increasingly stringent specifications for blast furnace coke.

A high proportion of coke manufactured in the U.K. is used in blast furnaces. Recent improvements in blast furnace and cupola technology have resulted in significant reductions in the coke required per tonne of iron produced. The injection of fuel oil into the furnace as a supplementary heat source, for example, is widespread, and it has been shown that pulverised coal can be used in place of fuel oil for this purpose. Reducing gases manufactured from coal by gasification using air, or oxygen enriched air, and steam can also be injected into furnace, further reducing the coke requirements. The downward trend in the coke rate (i.e. the amount of coke required per tonne of iron produced) can therefore be expected to continue, possibly until the end of the century.

Other smokeless fuels aimed at the domestic market are manufactured in significant quantities. Phurnacite is made by heat treating a pitch bound briquette in narrow ovens each holding around two tonnes, and is produced for closed appliances and roomheaters. Homefire is a fuel for open fires made by reducing the volatile content of crushed coal in a fluidised bed and hot-briquetting the resultant char. Coalite and Rexco also involve the carbonisation, in special retorts, of coal as smalls or sized coal respectively and are used on open fire and closed appliances. These processes were developed to attain a desired reactivity of the product by using particular types of coals. The reduced availability of the dry steam coal used originally for Phurnacite production has led to research on the use of other feedstocks, resulting in their partial replacement with blends of anthracite and more coking coals.

The older coke ovens, which were designed in a period when environmental considerations were of less account, give rise to fume and dust emissions during their operation. Substantial improvements are being achieved by modifications to existing machinery, a notable example being the development and general use of smokeless charging techniques. The liquid effluents from carbonisation processes present special difficulties. For example, those from coke ovens are treated for recovery of ammonia leaving a liquor containing phenols, cyanides, thioyanates, sulphides and other compounds. The main method of purification is by biological oxidation on site and/or discharge to sewers for further treatment by water authorities. Where these processes are not being used, discharges are made, with consent, to tidal estuaries or to the open sea.

4.5 Gasification

When supplies of natural gas are no longer sufficient to meet the premium markets for this fuel, the NCB expect substitution based on coal gasification to occur. Coal is gasified by reacting it with air or oxygen and, in most processes, steam. Three types of product gas can be obtained:

(i) low calorific value (lean) gas

This is produced when air is used as the main gasifying agent. The gas has a low energy content per unit volume, making transmission and distribution costs high. Applications would therefore be restricted to power station or bulk industrial use.

(ii) medium calorific value gas

This is produced using oxygen and steam as the gasifying agents. The absence of nitrogen (of low calorific value gas) makes this product suitable as a feedstock for synthesis processes or as a fuel gas.

(iii) substitute natural gas

SNG (methane) can be produced by upgrading the medium calorific value gas resulting from oxygen/steam gasification. The high energy content per unit volume makes storage and transmission costs comparatively low.

The optimum type, size and location of coal gasification plants for the U.K. is not yet clear. In the U.S.A., the concept of large SNG plants sited near the coal mine and producing about 8 million m^3 of SNG per day (equivalent in coal input to a 2000 MW power station) has become the accepted standard for the design studies so far carried out for commercial plant; low calorific value gas plants are proposed mainly in the context of reducing emissions from coal fired power generation systems. The existence of the gas grid in the U.K. will encourage the adoption of large central SNG plants, although the appropriate plant size may be smaller than proposed in the U.S.A.

Gasification plants can be designed to meet low levels of air pollutant emissions. The main atmospheric emissions arise from coal combustion for auxiliary power and process steam requirements, although these can be avoided by using clean gas for this purpose, at the expense of some loss in efficiency. By using chemical absorption systems, more than 99% of the sulphur can be retained. This can be recovered as elemental sulphur for ease of disposal, if required. Organic nitrogen in the coal is also recovered in the gas clean-up system, mainly as ammonia.

Several commercial gasification processes exist, the most important of which are:-

(i) the Lurgi fixed-bed gasifier.

(ii) the Winkler fluidised-bed gasifier.

(iii) the Koppers-Totzek entrained phase gasifier.

Various developments to improve these systems are taking place, including the British Gas 'slagging Lurgi' gasifier at Westfield in Scotland. This gasifier is based on the Lurgi process, but operates at a higher temperature to improve the gasification efficiency. The other major SNG gasification process development in the U.K. is the work on the COGAS system carried out by the NCB at CURL, Leatherhead. This process uses steam alone as the gasification agent, avoiding the cost of air separation plant for oxygen production. The slagging Lurgi and COGAS systems are the two main candidates for selection as a demonstration scale plant to be built by the U.S. Department of Energy.

4.6 Liquefaction

Both coal and oil were formed from organic remains, the product depending on the chemical compounds present at the time of formation and the geological processes acting on them. Compared with crude oil, coal has a deficiency of hydrogen; the hydrogen content is 6% maximum, whereas crude oil contains 10% to 14% hydrogen. This deficiency of hydrogen is responsible for coal being a solid, and any process to obtain a high yield of liquid fuels from coal therefore has to introduce additional hydrogen.

Most of the current developments programmes on coal liquefaction are concentrating on the extraction/ hydrogenation route. This involves dissolving coal in a solvent at about 400°C and reacting the solution with hydrogen gas at high pressure (c. 100 bar) in the presence of a catalyst. The techniques involved in the second stage can be regarded as an extension of modern oil refinery practice in which heavy oils are upgraded to light fractions by reaction with hydrogen.

The various extraction/hydrogenation liquefaction processes under investigation differ mainly in the technique employed for separating out the mineral matter, and the stage at which this occurs. The NCB have developed two such processes, one based on liquid solvent extraction and the other using a gas under supercritical conditions as the solvent. The NCB have submitted proposals to the Department of Energy for pilot plants on the scale of 1 t/h (coal input) to develop these processes further. The proposals have been endorsed by the Coal Industry Tripartite Group (Ref. 9) Government financial aid has been promised, and design studies have been initiated.

An alternative route to liquid fuels and chemicals from coal is synthesis. This involves manufacturing a medium calorific value gas (synthesis gas) from coal and then converting it to liquid products using a catalyst. The process is carried out commercially in South Africa at the Sasol plant. The economics of this route are generally considered to be less favourable because of its comparatively low thermal efficiency.

Another process which may form a component of liquefaction plant is pyrolysis. When coal is heated in the absence of oxygen, the hydrogen-rich species (the volatile matter) are removed as liquids and gases. Although this occurs in a conventional coke oven, the yield of liquids is low. By controlling the heating conditions carefully, up to 30% of the coal can be obtained as liquid products. The residue is a low volatile matter, coke-like material which can be used for further processing (e.g. as a boiler fuel or for gasification).

The liquid fuels produced by extraction/hydrogenation and pyrolysis processes have a high content of aromatic chemicals. The precautions against health hazards are referred to in Section 5.8.

The liquid fuels produced from synthesis gas by most processes differ from those produced by extraction/hydrogenation in having a low aromatic content. An advantage of the synthesis route is that the liquid fuels are virtually sulphur-free. This arises because of the high efficiency with which sulphur can be removed from the medium calorific value gas produced as the first stage of the process.

As with gasification, the main gaseous emissions from a liquefaction plant derive from combustion for power generation and process steam production. The emissions can be limited either by direct control, or by using some of the clean products for this purpose. Because liquefaction can be regarded as an extension of the existing technologies, the environmental problems are not expected to be different in nature from those of present oil refinery, carbonisation and power station activities.

5. ENVIRONMENTAL IMPLICATIONS OF COAL UTILISATION

5.1 Sulphur dioxide

When coal is burned, the sulphur is converted mainly to sulphur dioxide in the combustion gases, although smaller quantities of other sulphur oxides (e.g. SO_3) and sulphates formed from the mineral matter are also present.

Sulphur occurs in coal in two forms; organic sulphur exists as part of the coal substance, the remainder being mainly pyritic sulphur present in the mineral matter. The organic sulphur content of U.K. coals ranges from 0.4% to 1.5% with a mean of 0.8% (air dried basis). Pyritic sulphur averages 0.7% and ranges from 0.2% to more than 2.0% in exceptional cases.

The relatively high density of the pyrites makes possible some concentration of this form of sulphur in a particular coal fraction (the middlings) during washing, although coals differ in their suitability for this form of treatment. For example, some U.K. washed coal of 1.5% sulphur content could be separated into fractions comprising 85% of the coal with a sulphur content of 1.35% and fractions comprising 15% of the coal with a sulphur content of 2.35%. This method of washing is used to reduce the sulphur content of some of the coal prepared for the coking market, the high sulphur fraction obtained being sold for power generation.

A number of novel processes, based mainly on treatment by oxidation, have been proposed for reducing the sulphur content of coal, including the organic sulphur content. Research on these is being carried out in the U.S.A. and elsewhere, but is in an early stage.

The use of tall stacks by industry and by power stations together with the increased use of natural gas and electricity in domestic and commercial premises have reduced considerably the ground level concentrations of sulphur compounds, and the justification for further measures has not yet been established.

A 60% reduction in the mean urban sulphur dioxide concentration has occurred since 1960/61 and the trend is still downwards (see Table 5 and Ref. 11).

TABLE 5

Mean urban sulphur dioxide and smoke concentrations in the United Kingdom.

	Concentration (microgramme per m^3)	
Year	Sulphur dioxide	Smoke
1960/61	158	150
1965/66	116	96
1970/71	100	55
1975/76	71	32
1977/78	60	25

Atmospheric emissions from combustion processes can be classified (in relation to chimney height) as high level (from power stations), medium level (from industrial uses) and low level (from domestic). Projections of the change in the pattern of sulphur emissions in the U.K. to the end of the century from solid fuel usage have been made on the assumption that no special measures are taken to reduce these emissions and are given in Table 6.

TABLE 6

Estimated U.K. SO_2 emissions from solid fuel usage
(million tonnes per year)

Source	1977/78	2000
High level	2.11	2.41
Medium level	0.23	1.02
Low level	0.39	0.22
TOTAL	2.73	3.65

(The above figures assume that no sulphur retention measures are introduced. The increase in sulphur emissions therefore simply reflects the increase in coal consumption. On a national basis, this increase would be partially offset by displacement of oil of a similar sulphur content per unit of energy content.)

Sulphur emissions from sources in the U.K. have, for some time, been suspected of contributing to the acidity of precipitation in other countries, but the extent of this contribution has been shown by a recent OECD study (Ref. 5) to be considerably less than had previously been suggested. Evidence about the effect of transported pollutants on plant and fish life is uncertain and, in some respects, conflicting.

The sulphur contents of U.K. coals are generally lower than heavy fuel oils, both on a mass basis and per unit of energy content (see References 12, 14). Furthermore, some of the sulphur dioxide formed during the combustion of coal (generally accepted to be at least 10%, Ref. 13) is retained in the ash. Because coal competes mainly with heavy fuel oil for power stations and industry, a change towards coal in these markets would reduce sulphur emissions (for plant of equal thermal efficiency).

Current production plans indicate that the sulphur content of the output from existing pits will remain at the present level for the next 10 years. The sulphur content of classified and unclassified reserves at existing collieries plus Selby and Betws new mines is the same as that in present and planned output. However, the sulphur content in the classified portion of those reserves is slightly higher than in the unclassified reserves. To that extent, when these reserves are mined, there may be a slight increase in the average sulphur content, perhaps by 0.1 percentage point. However, this tendency will be offset by the relatively lower sulphur content expected in the output from new mines, at least until the end of the century.

In general, the sulphur contents of U.K. coals are c. 1.5%, in contrast to North American coals for which sulphur contents of 4% are common and 8% not unknown. Concern about the local effects of unrestricted combustion of such coals has been a central feature in formulating U.S. environmental control legislation. The proposed U.S. Environmental Protection Agency standards (Ref. 15) for new coal-fired plant larger than 25 MW(e) are that sulphur dioxide emission should not exceed 0.520 kg/GJ of heat input or 15% of the potential sulphur emissions, whichever is the lower. For low sulphur coals, the sulphur dioxide emission need not be reduced below 0.086 kg/GJ. The emission levels referred to are defined as maximum levels for 2 hour averages.

If it becomes desirable to reduce sulphur emissions from direct combustion, the two main techniques available are stack gas scrubbing and the use of fluidised bed combustion with a sulphur acceptor (see earlier). In either case, sulphur retention is easier and more practical for power station and large industrial applications than for small consumers.

The conversion of coal into synthetic liquid and gaseous fuels is accompanied by a reduction in sulphur content. In gasification processes, chemical absorption can remove over 99% of the sulphur from the gas. By separating out the mineral matter, liquefaction can produce a low sulphur fuel of similar chemical composition to coal; this has attracted interest in the U.S.A. as a power station fuel and is referred to as a solvent refined coal (SRC).

Hydrogenation further reduces the sulphur content and results in light low-sulphur liquid fuels. In the case of both gasification and liquefaction processes, the sulphur removed from the coal can be recovered in elemental form for disposal, if required.

The detailed implications of sulphur emissions are not yet fully understood. In particular, the health effects of the various forms of sulphur pollutants (e.g. sulphates) and the mechanisms of long-range transport are the subject of much debate. The NCB suggest that these uncertainties need to be resolved to provide a firmer basis for a view on the role of sulphur as a pollutant.

5.2 Oxides of nitrogen

Oxides of nitrogen (NO_x) are formed by the combustion of all fuels including petroleum products and natural gas. Although the detailed mechanisms of NO_x formation are not clear, it has been established that both the organic nitrogen in the fuel and the nitrogen in the air used for combustion can contribute.

The degree of NO_x formation depends on the combustion conditions being encouraged by a high nitrogen content in the fuel, a high air to fuel ratio high combustion temperature and rapid cooling of the gas. Typical levels of NO_x emission from different combustion plants are shown in Table 7. As illustrated in the table, NO_x emissions from coal-fired fluidised bed combustors are lower than from conventional coal-fired plant. This is because the comparatively low combustion temperature virtually eliminates NO_x formation from atmospheric nitrogen, so that the only source of NO_x is the organic nitrogen in the fuel. Operation of a fluidised bed combustor at elevated pressure further reduced NO_x emissions.

TABLE 7

Levels of NO_x formation

Fuel	Nitrogen in fuel (weight percent dmmf basis)	Type of plant	NO_x in flue gas (volume ppm, dry basis)
Coal	Range 1 to 2	PF power station boiler	200 to 1400
		Atmospheric pressure fluidised bed combustor	200 to 700
		Elevated pressure fluidised bed combustor	60 to 200
Fuel oil	0.2	Power station boiler	215
	1.0		425
	Range		110 to 800
Natural gas	Negligible	Power station boiler	50 to 1500

References 16, 17.

Concern over NO_x emissions has arisen mainly in the U.S.A. because of the creation of smogs in areas such as the Los Angeles basin, where special climate factors are important.

It is not established that emissions from the combustion of fossil fuels cause a significant hazard, except where special climatic factors operate, and not all of the states in the U.S.A. enforce the EPA recommended standards.

In the U.K., the climate is such that there appears, on present evidence, to be no cause for concern about the environmental effects of nitrogen oxides provided that current standards of dispersal of combustion gases (e.g. the U.K. "tall stack" policy) are maintained.

5.3 Ash

Coal, as mined, contains a considerable proportion of mineral matter, some of which is removed during coal preparation as colliery spoil. The mineral matter which remains with the coal is altered chemically by combustion and is then referred to as ash. The main categories of coal sold to the various market sectors are given in Table 8, together with the corresponding ash contents. The total production of ash in the U.K. is approximately 16 million tonnes per year.

TABLE 8

Ash production in the U.K. in 1975/76

	Coal production (million tonnes per year)	Ash content (%)	Ash production (million tonnes per year)
Power stations	73.1	17.1	12.5
Carbonisation	17.5	6.0	1.1
Industrial and miscellaneous	12.1	9.6	1.2
Domestic and manufactured fuels	15.0	4.7	0.7
Others including exports.	1.5	6.3	0.1
TOTAL	119.2	13.2	15.6

Reference 12

About half of the pulverised fuel ash (PFA) produced by power stations is sold as a construction material, the main markets being:

(i) As a load-bearing fill (3.6 million tonnes pa).

(ii) In building blocks and concrete building products (1.3 million tonnes p.a.).

(iii) In sintered lightweight aggregates (0.3 million tonnes p.a.).

(iv) In brickmaking (less than 0.1 million tonnes p.a

(v) In concrete and mortars (less than 0.1 million tonnes p.a.).

The remainder of the PFA and industrial ash is used mainly as a filling material (e.g. disused gravel pits or clay pits) or for dumping on low-lying land or at sea.

5.4 Dust and grit

In PF power stations, the method of particulate removal is by electrostatic precipitators. These remove all but approximately 0.7% of the particulate matter in the combustion gas; the efficiency decreases for particles in the submicron size range but is still in excess of 97%. The presence of sulphur oxides in the combustion gases assists the action of electrostatic precipitators, and a reduction in the sulphur content of the fuel will tend to impair their operation. For systems employing low sulphur coal (obtained, for example, by selective washing) or sulphur retention during combustion (e.g. fluidised bed combustion with the addition of a sulphur acceptor to the bed) other particulate removal systems, such as baghouse filters, may be necessary.

For industrial applications, the dust loading in the combustion gas is generally lower than for PF power stations. This arises partly because of the lower ash content of the washed coal used and partly because of the larger coal size which permits the removal of most of the ash from the grate on stoker systems. In general, industrial boilers and furnaces use cyclones or multicyclones to remove particulates. Typically, these capture about 70% of the particulates in the combustion gas, retaining most particles larger than 20 microns to 50 microns, depending on the cyclone design.

Estimates of the present and future emissions of particulates in the U.K. arising from solid fuel combustion are given in Table 9 (Ref. 12).

TABLE 9

Estimated U.K. particulate matter emissions from solid fuel usage

(million tonnes per year)

Source	1975/76	2000
High level	0.068	0.081
Medium level	0.036	0.078
Low level	0.323	0.176
TOTAL	0.427	0.335

5.5 Smoke

The problems of smoke emissions and smogs in the past have resulted mainly from the combustion of bituminous coal on domestic open fires in areas of high population density. Smoke emission occurs due to unburned volatile matter leaving the fire and can be avoided by burning smokeless fuels or by special combustion techniques (see Section 4.3).

The Clean Air Acts of 1956 and 1968, with their provision for the introduction of smokeless zones and regulations controlling industrial smoke emissions have been successful in virtually eliminating smoke emission problems (Fig. 4). Table 5 shows the average smoke concentration in urban areas of the U.K. in 1977/78 to be one eighth of that in 1960/61 and that average concentrations are still falling.

The present domestic smoke emission standards required by the Department of the Environment can be summarised as follows:

(i) Smoke emissions from manufactured "smokeless" fuels should not exceed 5 g/h (when tested to BS 3841).

(ii) Smoke emissions from smoke reducing appliances should not exceed the smoke emission from a "smokeless fuel" by more than 0.1 g/h per Btu/h of total output (when tested according to the recommendations of PD 6434).

FIGURE 4 Sheffield: Before and after smoke control
(Courtesy of Sheffield Publicity Department)

The coal industry has actively assisted the progress achieved over the past two decades through the introduction of smokeless fuels and appliances. The NCB operate, at the Coal Research Establishment, fuel and appliance testing laboratories to DOE standards for approval purposes. This continues to ensure that the present high standards are maintained.

5.6 Trace elements

Concern has been expressed that coal utilisation may introduce hazardous trace elements into the environment. The elements are present mainly in the mineral matter associated with the coal and can enter the environment in two ways:

(i) By leaching from ash or minestone tips.

(ii) As particulates emitted to the atmosphere as a result of combustion.

The ash from coal contains only a small amount of soluble material, and that which is leached from dumps in general dilutes rapidly in underground and surface water (Ref. 19).

Small quantities of trace elements are emitted to the atmosphere with combustion gases as sub-micron particles and vapours. However, the "tall stacks" policy dilutes and disperses emissions, including those of trace elements.

Some of the main elements of concern are listed in Table 10, together with the form in which the elements are removed from the plant.

TABLE 10

Trace element concentrations

Element	National average concentration in the coal (ppm)	Range of Coalfield Averages (ppm)	Occurrence of compounds containing element
Arsenic	18	4-40	Mainly in fly ash
Cadmium	0.4	0.3-0.6	Mainly in fly ash
Mercury	0.5	0.4-0.6	Vapour in combustion gases
Lead	38	28-60	Mainly in fly ash
Vanadium	76	38-134	Slag and fly ash
Fluorine	114	98-130	Vapour in combustion gases

References 12, 20

Recent work by the U.K. Central Electricity Generating Board has shown that the concentration of trace elements in the environment resulting from the emission of particles from coal-fired power stations are about two orders of magnitude lower than those prevalent in U.K. urban areas resulting from a combination of natural and other man-made sources.

5.7 Carbon dioxide

Carbon dioxide is produced when any organic fuel (coal, oil, natural gas, peat, wood, etc.) is burned. It seems to be now well established that the concentration of carbon dioxide in the atmosphere is increasing. The concentration by volume in 1974 was 330 ppm, compared with 313 ppm in 1958 and an estimated value of 290 ppm in 1870 (Ref. 21). Only part of this increase can be attributed to fossil fuel combustion; it has been estimated that the carbon dioxide produced by burning wood during deforestation is of the same order of magnitude as that produced by the combustion of fossil fuel in this century.

The continued increase in carbon dioxide concentration has led to speculation that significant changes in climate on a global scale may result in the future. This could occur by interference with the balance between the radiation received from the sun and that emitted by the earth. The characteristics of carbon dioxide are such as to absorb the low frequency radiation emitted by the earth but not to have a corresponding effect on the radiation from the sun. This radiation trap is known as the "greenhouse effect".

Considerable uncertainty surrounds predictions of the extent to which the carbon dioxide concentration will continue to increase, as this depends on the transfer of carbon between the atmosphere, vegetation and the oceans. Further uncertainty surrounds the effect of carbon dioxide on the climate; at present no climatic change above the "noise level" of natural variations has been detected.

In view of the complexities of climatic changes and the carbon cycle, much work has yet to be done before sufficient is known to form a view on the extent of the "greenhouse effect". The NCB therefore maintain a watching brief on research in this area.

5.8 Hycrocarbon compounds

Careful design and operating procedures have to be followed in plants handling hydrocarbon compounds in order to prevent health hazards. In particular, the coal industry has been active in developing and improving atmospheric monitoring devices to ensure a safer working environment in coking and related plants. The introduction of coal liquefaction and gasification plants in the future is not expected to lead to problems of a different nature to those in existing plants.

Emissions of hazardous compound through consumption of liquid fuels is also an area of concern. In motor transport applications, for example, unburned fuel may be emitted to the atmosphere in exhaust fumes. Petrols with a high content of aromatic compounds, although undesirable from the viewpoint of hydrocarbon emissions, require less lead additives for a given octane rating. These problems are similar whether oil or coal derived fuels are used, and are being investigated by the oil industry.

5.9 Waste heat

Almost all energy consumption ultimately results in heat being transferred to the environment. The dispersed nature of some sectors (e.g. domestic and transport) is normally such as to have little effect. Some changes in the local environment can occur as a result of large energy conversion processes (e.g. power stations).

The amount of waste heat produced by a process is related directly to the thermal efficiency. A modern power station has a thermal efficiency of 35%, and therefore discards 65% of the heat input to the environment. An oil refinery, by contrast, has a thermal efficiency of, typically, 90% and only discards 10% of the heat input. The conversion of coal to SNG and liquid fuels has a thermal efficiency of 60% to 70%, so that 30% to 40% of the heat input is rejected to the environment.

Any method of improving efficiency or reducing energy consumption will reduce both emissions and the amount of heat released to the environment from large plants. Combined heat and power systems, for example, utilise some of the 65% of the heat input discarded from a power station for district heating. However, these suffer from the disadvantages of reducing the amount of electricity generated and requiring a considerable investment for the hot water distribution system.

An alternative approach is to take advantage of the effect of waste heat on the environment. For example, waste heat from power stations is being used to heat greenhouses and to encourage fish growth in rivers or estuaries.

5.10 Water and effluents

The water requirements for large energy conversion processes can be considerable and can influence their location. Typical water requirements for several processes are given in Table 11 (Ref. 22). The coal conversion processes, although requiring more water than an oil refinery, require considerably less than a power station.

TABLE 11

Water requirements of conversion processes

Process	Water requirements (litres per gigajoule of output)
Oil refinery	50 to 100
Coal liquefaction	c. 220
SNG from coal	130 to 260
Coal-fired power station	c. 900

In a power station, however, water is used mainly for cooling and is discharged warm but uncontaminated. In coal conversion processes (including coke manufacture), water is also needed for scrubbing and process steam duties. Some of this water is recovered contaminated with hazardous compounds, including phenols, cyanides, thiocyanates, thiosulphates and sulphides. These effluents have to be treated by biological oxidation before discharge to the environment.

5.11 Land requirements

The land requirements for a conventional PF power station operating on base load are about 25 to 60 hectares per million tonne of annual coal input; the estimated land requirements for the coal conversion processes under development are expected to be somewhat lower (Ref. 22). This compares with current requirements of approximately 200 hectares of operational land per annual million tonne of coal produced by deep-mining and 133 hectares per annual million tonne for opencast. The land requirements for the large-scale conversion and use of coal are, therefore, considerably less than for coal production.

REFERENCES

1. "Digest of United Kingdom Energy Statistics 1978", Department of Energy, London, HMSO.
2. "Energy conservation in the United Kingdom; achievements, aims and options", National Economic Development Office, London: HMSO 1974.
3. NCB Press Release, January 22, 1979. Statement by the Chairman, Sir Derek Ezra, to the Commission on Energy and the Environment.
4. D.J. Moore and A.G. Robins, "Experimental and theoretical investigations into chimney emissions", Proc. Inst. Mech. Engrs., Vol. 189, 4/75, 1977.
5. "Long range transport of air pollutants; final report of a co-operative technical programme prepared for the OECD by the Norwegian Institute for Air Research", OECD Paris, 1977.
6. "Experimental Husbandry", Vols 4,6, 7, Ministry of Agriculture, Fisheries and Food, 1951.
7. "Plan for Coal", published by the NCB, Hobart House, Grosvenor Place, London, June 1974.
8. "The economics of electricity from coal, nuclear and regenerable sources", Economic Assessment Service, NCB (IEA Services) Ltd., Lower Grosvenor Place, London, Draft report dated December, 1978.
9. "Coal Technology; Future developments in Conversion, Utilisation and Unconventional Mining in the United Kingdom", Department of Energy, Thames House South, Millbank, London.
10. Centre piece No. 13, Nov. 1975, published by the Solid Fuel Advisory Service, Hobart House, Grosvenor Place, London.
11. "Estimates of smoke and SO_2 pollution from fuel combustion in the U.K.", Warren Spring Laboratory, 1977, published in "Clean Air", Winter, 1977.
12. "Study into the emission of air pollutants coming from the use of coal within the United Kingdom, Part 1", prepared for the Environment and Consumer Protection Service of the Commission of the European Communities by the NCB, Hobart House,Grosvenor Place, London, 1978.

13. Memorandum on Chimney Heights, "Chimney Heights", 2nd Ed., Ministry of Housing and Local Government, 1967.
14. Cooper, J.R. and Rose, J.W., "Technical Data on Fuel", 7th Edition 1977, The British National Committee, World Energy Conference.
15. JAPCA, Vol. 28, No. 11, Nov. 1978, p 1148.
16. "Combustion Generated Pollution", Science Research Council, July 1976.
17. H.R. Hoy and H.B. Locke, Paper to the Swedish Institute of Engineers, Stockholm, Oct. 1972.
18. "U.S. EPA standards of performance for new station resources", Federal Register, Vol. 36: 24876, Dec. 1971.
19. "The disposal of pulverised fuel ash in water supply catchment areas", J. Brown et al, Water Research, Vol. 10, 1976.
20. Richards, D.J.W., Jones, W.S., and Laxton, J.W., "Monitoring and control of flue gases", CEGB Research, Aug. 1977.
21. "Carbon dioxide and the greenhouse effect; an unresolved problem", a report by the Technical Information Service, NCB (IEA Services) Ltd., Lower Grosvenor Place, London, April 1978.
22. "Liquid fuels from coal", published by the National Coal Board (Coal Research Establishment, Stoke Orchard, Cheltenham, Glos.) August 1978.
23. Coal Industry Examination, Interim Report, June 1974, Department of Energy, Thames House South, Millbank, London.
24. Coal Industry Examination, Final Report 1974, Department of Energy, Thames House South, Millbank, London.
25. "Coal for the Future; Progress with Plan for Coal and prospects to the year 2000", Department of Energy, Thames House South, Millbank, London.